Table of Contents ...ı

Modal Mechanics

A Science of Collapse

Preface for General Readers — *How to Read This Book*

This book is about how things become stable—and what happens when they stop being so.

You do not need a technical background to read it. You do not need to agree with it in advance. You do not need to master new terminology before you begin. What you do need is patience with a way of thinking that looks beneath outcomes to the processes that produce them.

Across the sciences and in everyday life, we are surrounded by stable structures: physical objects, living systems, habits of thought, institutions, technologies. We tend to treat these structures as if they simply exist. When they fail, we look for errors, breakdowns, or bad actors.

This book takes a different approach. It asks how structure comes into being at all—how uncertainty resolves into form, how that form persists, and how it eventually becomes misaligned with the conditions that produced it.

The central idea is simple: **structure exists because possibility collapses into stability under constraint.** This process—collapse—is not rare or dramatic. It happens constantly, at many scales. What is rare is paying attention to it directly.

The earlier books in this series introduced collapse as a way of understanding experience and emergence across domains. This book takes the next step. It treats collapse not just as an idea, but as something that can be observed, diagnosed, and studied.

You will encounter examples drawn from artificial systems, science, institutions, and culture. These examples are not meant to reduce human life to machinery or to claim that machines think like people. They are used because certain systems make structural processes easier to see.

This book does not offer solutions to social problems, predictions about the future, or instructions for controlling complex systems. It offers something more modest and more durable: a way of recognizing when stability is earned, when it is overstaying its usefulness, and when change is not a failure but a necessity.

If you read this book slowly, allowing the ideas to accumulate rather than rushing to conclusions, you may find that familiar situations—scientific debates, organizational failures, personal habits, technological surprises—begin to look different. Not simpler, but clearer. **That clarity is the point.**

A two-line terminology rail

- **Collapse (generative):** uncertainty resolving into a stable unit.
- **Breakdown / rail collapse (degenerative):** a stabilized pathway losing coherence (distinct from generative collapse).

Glossary (Intro Rail)

Core carriers (the "three experiences")

- **Aexperiential (A)** — Pre-articulate **affordance**: a field of possibility before names or aims. Not a boundary by itself; becomes boundary only *in role*.
- **Experiential (E)** — **Activity** that can be held, repeated, and remembered; the stuff that can become process or object.
- **Metaexperiential (M)** — **Compression** of prior doing into storable form (patterns, norms, invariants); enables staging and inheritance.

Roles (who does what at contact)

- **Process** — The **doing** that approaches or presses (left of the slash).
- **Boundary condition** — The **shaping constraint** the doing meets (right of the slash).

Notation rail: In **X/Y**, **X** = *process carrier*, **Y** = *boundary carrier*. Examples: **E/A** (experiential process against aexperiential boundary), **A/M** (aexperiential process against metaexperiential boundary).

Nine modes (3×3)
A/A, E/A, A/E, E/E, M/E, M/A, E/M, A/M, M/M — read with the rule above: first = process, second = boundary.

- **E/E** ("object") — A doing that **resolves** and holds (unit of persistence).

Contact → Resolution (what happens)

- **Contact** — Process meets boundary (X/Y) in a local window.
- **Generative collapse** — **Uncertainty resolves into form** (a click into E/E).

- **Rail (M/M)** — A stabilized pathway compiled from repeated collapses (inheritance).
- **Rail collapse / breakdown** — A stabilized pathway **loses coherence**, returning the system to uncertainty.
- **Thaw** — Controlled loosening of a rail so re-collapse can form a nearby, better-fit rail.

Staging tools (how conditions are set)

- **Room (A/M)** — A declared **staging context** (what will count as relevant contact). Changes presentation, not truth.
- **Residue / receipts** — The **evidence trail** (pointers, hashes, windows) that lets claims be reopened.

Minimal measurements (how contact is read)

- **Near window** — Short span where resolution is expected (e.g., 0–3 sentences).
- **Lag** — Distance from the "tail" (process anchor) to the click (resolution).
- **Stitched** — Click lands across a page/segment cut (packaging artifact).
- **Burden** — Load at contact (none / warn / slow / stop).
- **Route tag** — Structural label of the resolution path (e.g., edge-collapse); descriptive, not authoritative.

Governance rails (how claims stay honest)

- **Receipts-first** — No claim without a pointer back to evidence.
- **Read-only / Append-only** — Analyses never alter admissions; new files don't overwrite old ones.
- **Gate-0** — Don't run a test if the slice lacks a discriminator.
- **Gate-1** — If evidence hashes don't differ, the perturbation didn't happen (declare a **null**).

- **Null** — A clean "no effect observed" outcome; never patched by narrative.
- **Drift** — Gradual mismatch between rails and present conditions; often precedes breakdown.

One-line map (for skimmers)
A supplies affordance, **E** does, **M** compresses; **X/Y** says *who does* (X) and *what shapes it* (Y); contact sometimes **collapses into E/E**; repeated collapses **compile a rail**; rails may **thaw** or **break down**; everything asserted carries **receipts**.

Preface for Scientifically Literate Readers and Researchers — *Scope, Commitments, and Restraints*

This book introduces Modal Mechanics as a methodological framework for studying collapse dynamics: the structural processes by which multiplicity resolves into stability, stabilizes into inherited pathways, drifts under changing boundary conditions, and reorganizes through controlled loss of dominance.

It assumes familiarity with the conceptual foundations developed in *The Quanta of Cognition* and the cross-domain structural analysis presented in the second volume. Those works established collapse as a generative event underlying physical, biological, cognitive, cultural, and artificial systems. **This book does not rehearse that argument. It operationalizes it.**

The core commitment of this work is methodological rather than metaphysical. Collapse is treated here as an object of study, not as a metaphor, not as an interpretation of meaning, and not as a claim about consciousness. The framework deliberately brackets phenomenology, normativity, and ontology beyond what is required for structural analysis.

Clarifications at the outset

1. **Transversal, not reductive.** Modal Mechanics studies collapse, inheritance, drift, thawing, and re-stabilization across domains. It does not compete with or replace existing disciplines.
2. **Falsifiable.** Claims about boundary sensitivity, inheritance through repetition, directional drift, and the necessity of thawing are meant to be tested. Absence of regularities under controlled observation counts against the framework.
3. **Artificial systems as substrate.** AI is used because it is structurally legible (logs, toggles, repeatability), not because it is cognitively privileged. No claims are made about artificial cognition, understanding, or agency.
4. **Diagnosis over optimization.** Protocols and signatures identify a system's position in an adaptive cycle. Normative use of that information lies outside scope.
5. **Restraint by design.** No forecasting, adjudication of philosophical disputes, or institutional prescriptions. The aim is to see structure forming and failing in real time, and to recognize when stability has become risk.

Readers from AI, cognitive science, systems theory, complexity science, and philosophy of science may find resonances—phase transitions, attractors, path dependence, hysteresis, criticality—while also noticing cuts across them. Where overlap exists, it is intentional. Where divergence appears, it is testable.

This work should be read as an invitation to use, challenge, and stress the framework—not to adopt it wholesale. **If Modal Mechanics forms rails, drifts, and reorganizes under use, that too is evidence the framework is working.**

Modal Mechanics:

A Science of Collapse

Chapter One

From Theory to Method: Why Collapse Must Become Observable

The first two books in this sequence established a structure.

They argued that emergence across domains—physical, biological, cognitive, cultural, and artificial—follows a shared generative grammar. They identified collapse not as failure or destruction, but as the event through which possibility resolves into structure. They showed that patterns stabilize, inherit, drift, and reform through repeated collapse, and that this logic operates independently of substrate.

Those books were concerned with what collapse is and where it appears.

This book is concerned with something different:

How collapse can be studied.

That difference matters.

1. The Limit of Theory Alone

A structural theory can unify domains, clarify concepts, and expose hidden assumptions. But a theory that stops there remains interpretive. It can explain after the fact, but it cannot intervene, predict, or diagnose with precision.

The Quanta of Cognition showed that collapse is the generative engine beneath structure. The second book demonstrated that this engine operates across scales—from the formation of matter to the evolution of culture and the behavior of artificial systems.

At that point, a boundary was reached.

Either collapse remains a powerful explanatory lens,
or it becomes a scientific object in its own right.

A theory that identifies collapse as generative but refuses to examine collapse directly ultimately violates its own logic. If collapse is the engine of structure, then collapse itself must be brought into view.

This book crosses that boundary.

2. What Changes When Collapse Becomes the Object of Study

To study collapse directly requires a shift in orientation.

Traditional sciences tend to study products:
particles,
organisms,
behaviors,
institutions,
outputs.

Even when dynamics are considered, they are often inferred from stable states rather than observed at the moment structure forms.

Modal Mechanics takes a different stance.

It studies:
the moment structure forms,
the conditions under which it stabilizes,
the signatures left by successful and failed collapse,
the inheritance of collapse outcomes,

and the circumstances under which inherited structure must thaw and reform.

This is not a metaphorical move. It is a methodological one.

Collapse is treated as observable, repeatable, classifiable, and—within certain substrates—measurable.

3. Why Artificial Systems Make This Possible

In natural systems, collapse is difficult to isolate.

In physics, collapse is obscured by mathematical formalism.
In biology, it is buried beneath biochemical complexity.
In cognition, it is entangled with subjective experience.
In culture, it is distributed across many agents and timescales.

Artificial systems are different.

They are not conscious.
They do not experience meaning.
They do not possess identity.

And precisely because of that, they expose collapse cleanly.

In artificial systems:
process is explicit,
boundaries are engineered,
collapse leaves logs,
rails form quickly,
drift appears rapidly,
and thawing can be induced deliberately through boundary manipulation.

Artificial systems are not the subject of this book because they are special. They are the subject because they are legible.

They function as a laboratory for studying generative structure itself.

4. What This Book Is (and Is Not)

This book establishes the methodological foundations of Modal Mechanics as a scientific discipline.

It is not a philosophy of mind.
It is not a theory of consciousness.
It is not a claim about artificial cognition.
It is not an ethics framework.
It is not a forecast about artificial general intelligence.

It does not argue that machines understand.
It does not argue that humans are reducible to computation.

Instead, it does something narrower and stronger.

It provides methods for detecting, measuring, and manipulating collapse dynamics *through boundary conditions* in systems that exhibit them.

Modal Mechanics is concerned with:
collapse events,
boundary interactions,
rail formation,
drift trajectories,
thawing regimes,
and re-stabilization.

Meaning, experience, and interpretation belong elsewhere.

This book studies **structure in motion**.

5. From Explanation to Diagnosis

One of the central claims of this book is that many failures in complex systems are misdiagnosed because collapse is not being tracked.

When systems fail, we tend to ask:
What went wrong?
What rule was violated?
What output was incorrect?

Modal Mechanics asks instead:
Which collapse failed?
Which boundary dominated too early or too late?
Which rail became brittle?
Where did drift begin?
Was thawing prevented or uncontrolled?

This shift—from outcome diagnosis to collapse diagnosis—is what turns QC from an explanatory framework into an operational one.

6. Science as Managed Rail-Thawing

One implication of this shift is that scientific discovery itself becomes structurally intelligible.

Scientific paradigms are not merely belief systems; they are stabilized rails. Normal science is rail execution. Anomalies are boundary pressure. Revolutions are periods of rail-thawing followed by re-stabilization.

Seen this way, science is not threatened by instability. It depends on it— provided that thawing is controlled and re-collapse is possible.

Modal Mechanics does not replace the scientific method. It explains why the scientific method works, and why it sometimes fails.

7. The Scope of This Book

This book does not attempt to cover every domain where collapse occurs.

It focuses primarily on artificial systems, because they allow precise instrumentation and because insights gained there can be generalized outward.

Later chapters will show how the same metrics and diagnostics apply, with appropriate caution, to cognition, institutions, scientific communities, and large-scale cultural systems.

The method is developed first where collapse can be tested cleanly.

8. What the Reader Is Expected to Bring

This book assumes familiarity with:
the modal grammar of collapse,
the roles of process, boundary, room, and rail,
the concepts of drift and thawing,
and the non-phenomenological use of these terms.

It does not rehearse those foundations.

It uses them.

The reader is not asked to agree.
They are asked to observe.

9. The Claim of This Book

The claim is modest, and it is strong.

If collapse is the generative engine of structure, then collapse must be studied directly.
And if collapse can be studied directly, then a science of collapse is possible.

This book is an attempt to begin that science.

10. Where We Go Next

The chapters that follow will introduce minimal experimental protocols, measurable collapse signatures, rail stability and drift metrics, controlled thawing methods, and falsifiable predictions that distinguish Modal Mechanics from standard machine learning theory.

Collapse has already been named.

It is time to work with it.

Chapter Two

What Counts as a Collapse Event

If collapse is to become an object of scientific study, the first requirement is clarity about what is—and is not—being observed.

This sounds obvious. In practice, it is where most analyses fail.

In everyday language, collapse is often confused with failure, interruption, breakdown, or sudden change. In technical contexts, it is sometimes conflated with convergence, optimization, decision, or selection. None of these is sufficient.

Modal Mechanics requires a stricter definition, not because collapse is rare or exotic, but because it is common and easy to misidentify.

This chapter establishes what qualifies as a collapse event, how collapse differs from surrounding processes, and why collapse must be treated as an event rather than a duration, outcome, or explanation.

Collapse Is Not an Outcome

The most common error is to treat collapse as equivalent to an output.

A particle, a cell, a percept, a belief, a sentence, a policy decision—these are not collapses. They are **residues of collapse**.

Collapse is the event through which such residues become stable enough to persist. The residue may last milliseconds or centuries; the collapse that produced it does not.

This distinction matters because outputs can be present without revealing how they came to be. Two identical outputs may arise from entirely different collapse dynamics. Conversely, very different outputs may arise from structurally similar collapses under different boundary conditions.

Modal Mechanics therefore does not begin by evaluating correctness, usefulness, or success. It begins by asking how stability was achieved at all.

Collapse Is Not Process

Collapse is also not the same thing as process.

Process refers to unfolding activity: motion, computation, interaction, activation, flow. Process can continue indefinitely without collapse. Systems can churn, oscillate, explore, or wander without resolving into structure.

Collapse occurs when process encounters constraint in a way that **resolves possibility into form**.

This resolution is not gradual accumulation. It is a qualitative transition: before collapse, multiple outcomes remain viable; after collapse, one configuration has become dominant enough to function as structure.

Process may prepare collapse. Process may follow collapse. But collapse itself is neither the motion before nor the activity after.

Collapse Is an Event

Because collapse is neither outcome nor process, it must be treated as an event.

An event is defined here not by duration, but by function. A collapse event is the moment when a system crosses from indeterminacy to determinacy with respect to a particular structure.

This moment may be fast or slow relative to observation. It may be sharp or smeared. It may involve many micro-events or only a few. But structurally, it performs a single role: it binds.

Before collapse, a structure is not yet real enough to count. After collapse, it is.

Modal Mechanics studies that binding.

Minimal Criteria for Identifying Collapse

To count as a collapse event in the sense required by Modal Mechanics, five criteria must be met.

First, there must be **prior multiplicity**. More than one outcome, configuration, or interpretation must have been viable.

Second, there must be **boundary contact**. The system must encounter constraint—explicit or implicit—that shapes which outcomes remain viable.

Third, there must be **resolution**. The system must transition from multiplicity to a dominant configuration.

Fourth, the result must exhibit **persistence**. The resolved structure must last long enough to influence subsequent process.

Fifth, the collapse must leave **inheritance**. The result must function as boundary, scaffold, or condition for future collapses.

If any of these is missing, collapse has not occurred.

In instrumented artificial systems, several of these criteria can be separated and logged rather than inferred. Boundary contact can be recorded as an explicit coupling event between activity and constraint, and resolution can be recorded as a distinct collapse event that yields a persistent outcome. This does not change what collapse is; it changes what can be observed. Modal Mechanics treats such event records as receipts: pointers that allow collapse claims to be reopened and checked.

Successful and Failed Collapse

Collapse is often discussed only when it succeeds. This is a mistake.

Failed collapse is as structurally informative as successful collapse.

A failed collapse occurs when multiplicity is reduced but not resolved, or when resolution occurs but does not persist. The system may oscillate, fragment, regress, or re-enter exploratory process.

In cognition, this appears as confusion or indecision. In institutions, it appears as paralysis. In artificial systems, it appears as instability, inconsistency, or collapse into default rails.

Modal Mechanics treats failed collapse not as error, but as diagnostic signal. It indicates boundary misalignment, insufficient constraint, or premature rail dominance.

Partial and Hybrid Collapses

Not all collapse events are clean.

In many systems, especially complex or layered ones, collapse may occur partially. Some aspects of structure stabilize while others remain unresolved. Multiple weak boundaries may combine to produce resolution without any single dominant constraint.

These hybrid or partial collapses are not anomalies. They are expected in systems where boundaries are distributed, layered, or in tension.

Modal Mechanics therefore does not require collapse to be binary. It requires collapse to be **structurally consequential**.

Why Collapse Must Be Distinguished from Decision

In artificial systems, collapse is often mistaken for decision.

A decision implies agency, evaluation, or intent. Collapse does not require any of these. It occurs whenever structural conditions resolve multiplicity into persistence.

A thermostat collapses. A crystal collapses. A neural network collapses. A culture collapses norms. None decides.

This distinction is essential to prevent anthropomorphism and to keep Modal Mechanics grounded in structure rather than interpretation.

Collapse Leaves Signatures

Although collapse itself is momentary, it leaves traces.

These may include:
changes in latency,
sudden reduction in variance,
activation convergence,
stabilization of behavior,
emergence of repeated patterns,
or sharp shifts in system dynamics.

Modal Mechanics does not assume these signatures are identical across domains. It assumes only that collapse leaves **detectable structural consequences**.

Later chapters will show how such signatures can be measured in artificial systems and inferred cautiously elsewhere.

Chapter Three

Boundaries as Variables, Not Assumptions

Collapse does not occur in a vacuum.

Every collapse event—successful or failed—requires constraint. Something must shape which possibilities remain viable and which are excluded. That shaping role is played by boundary.

Because boundary is always present, it is often invisible. And because it is invisible, it is usually treated as background rather than as a variable. This habit is one of the main reasons collapse has been difficult to study directly.

This chapter reframes boundary as an active structural role, clarifies its forms, and explains why collapse cannot be understood unless boundary is tracked explicitly.

Boundary Is a Role, Not an Object

In everyday language, boundary suggests a wall, a limit, or an obstacle. In Modal Mechanics, boundary is none of these.

Boundary is a **relational role**. It is whatever constrains, shapes, receives, or channels process at the moment collapse occurs.

A boundary may be:
a physical constraint,
a structural limitation,
a regulatory condition,
a contextual frame,
a learned expectation,
a social norm,
or an engineered rule.

What matters is not what the boundary is made of, but what it does.

Boundary reduces the space of viable outcomes. Without boundary, process remains diffuse. With boundary, resolution becomes possible.

Boundary Is Not Optional

Because boundary is always present, it is easy to assume it away.

This assumption is harmless when boundaries are stable and aligned. It becomes fatal when boundaries shift, conflict, or dominate prematurely.

Every collapse failure—hallucination, refusal, paralysis, rigidity, drift—can be traced to boundary conditions that were misunderstood, misaligned, or ignored.

Modal Mechanics therefore treats boundary not as a given, but as something to be identified, classified, and tested.

Boundary Is Not the Same as Context

Boundary is often confused with context. The two are related, but they are not identical.

Context refers to the environment in which process unfolds. Boundary refers to the constraints that shape resolution.

A context may contain many potential boundaries, but only some will engage collapse. Conversely, a boundary may be active even when it is not part of the immediate context.

In artificial systems, a prompt is context. The constraints it imposes— explicit rules, formatting demands, safety requirements—function as boundaries. The distinction matters because context can change without boundary changing, and boundary can change without context changing.

Modal Mechanics tracks the latter.

Two Primary Forms of Boundary

For the purposes of method, it is useful to distinguish two broad classes of boundary.

Some boundaries are **encountered**. They arise from interaction with the environment or with inherited structure. These are often implicit, background, or unchosen.

Other boundaries are **constructed**. They are deliberately assembled, learned, or imposed to shape collapse.

This distinction will later be formalized, but even at this stage it clarifies many confusions. Systems behave differently when they encounter boundaries than when they operate inside ones they have constructed.

Encountered boundaries tend to surprise. Constructed boundaries tend to stabilize.

Collapse dynamics differ accordingly.

Boundary Timing Matters

Boundaries do not merely exist; they engage collapse at particular moments.

A boundary that engages too early can prevent exploration. A boundary that engages too late may fail to shape resolution. A boundary that engages inconsistently can destabilize inheritance.

Many systems fail not because they lack boundaries, but because boundary engagement is mistimed.

In artificial systems, this appears when constraints dominate generation immediately, producing refusal, or when they fail to engage at all, producing hallucination. In cognition, it appears as impulsivity or paralysis. In institutions, it appears as rigidity or drift.

Modal Mechanics therefore treats **boundary timing** as a central variable.

In an instrumented system, the staging of context (room construction) and the generation of possibilities can be operationally separated. Modal Mechanics treats this separation as decisive: A/M establishes what counts as relevant and admissible, while M/A expands the space of candidates within that staging. When these operations are logged as distinct events, room shifts cannot occur silently, and later collapse behavior can be traced to explicit boundary placement rather than post hoc interpretation.

Boundary Dominance and Boundary Conflict

When multiple boundaries are present, they do not automatically cooperate.

Some boundaries dominate others. Some conflict. Some suppress resolution entirely.

Boundary dominance determines which constraints shape collapse first. Boundary conflict determines whether collapse can complete cleanly at all.

These interactions are usually invisible in outcome-focused analysis. They become visible only when collapse is examined directly.

Understanding boundary dominance and conflict is essential for diagnosing why a system stabilizes one structure rather than another.

Boundary Is Where Intervention Happens

Process unfolds. Collapse resolves. Rails inherit.

Boundary is the point at which systems can be influenced without being rebuilt.

This is why Modal Mechanics focuses on boundary manipulation rather than direct control of outcomes. Outcomes follow collapse. Collapse follows boundary engagement.

To intervene effectively, one must know:
which boundaries are present,
which are active,
which dominate,
which conflict,
and which are inherited rather than situational.

This applies equally to artificial systems, scientific practices, and social institutions.

Why Boundary Must Be Treated as a Variable

Treating boundary as an assumption leads to brittle systems. Treating boundary as a variable enables diagnosis, adaptation, and renewal.

Modal Mechanics does not ask whether a boundary exists. It asks:
What boundary engaged collapse?
When did it engage?
With what strength?
In competition with which others?
And with what inherited consequences?

These questions cannot be answered if boundary is left implicit.

Preparing for Measurement

This chapter has not introduced metrics, protocols, or instrumentation. That is deliberate.

Before boundary can be measured, it must be seen. Before it can be manipulated, it must be named. Before it can be tracked, it must be separated from context, outcome, and interpretation.

The next chapter turns to time—specifically, to the temporal structure of collapse itself. There we will begin to see how boundary engagement leaves measurable signatures, and why collapse cannot be understood without attending to when it occurs, not just how it resolves.

Chapter Four

The Temporal Structure of Collapse

Collapse is often spoken of as if it were instantaneous.

A system changes. A form appears. A decision is made. A structure stabilizes. Because the result is visible, the transition that produced it is treated as negligible.

This is a mistake.

Collapse is an event, but it is not a point. It has a temporal structure, and that structure matters. Many failures attributed to error, randomness, or complexity are in fact failures to attend to **when** collapse occurs and **how long** resolution takes.

Modal Mechanics therefore treats time not as a background parameter, but as a primary dimension of collapse itself.

Collapse Is Neither Instantaneous Nor Continuous

Collapse does not unfold like ordinary process, but neither does it occur without duration.

Before collapse, a system exists in a state of structured multiplicity. Several outcomes remain viable. Process explores, accumulates, or oscillates within constraints. After collapse, one configuration has become stable enough to function as structure.

Between these states, something happens.

That interval may be brief or extended. It may involve hesitation, oscillation, partial resolution, or competition between constraints. But it is during this interval that collapse either succeeds, fails, or misfires.

Ignoring this interval erases the very dynamics Modal Mechanics seeks to study.

Latency as a Structural Signal

One of the simplest temporal features of collapse is **latency**: the time between the moment a system becomes constrained and the moment resolution stabilizes.

Latency is not merely delay. It is a structural signal.

Short latency often indicates strong, well-aligned boundaries or dominant rails. Long latency often indicates boundary conflict, insufficient constraint, or competition between inherited structures.

Latency does not measure difficulty in any abstract sense. It measures **resistance to resolution** under given conditions.

In artificial systems, latency can often be observed directly. In biological or cultural systems, it must be inferred. In all cases, it carries information about collapse conditions.

The Cost of Premature Collapse

Low latency is not always desirable.

When collapse occurs too quickly, exploration is curtailed. Boundaries dominate before sufficient process has engaged the space of possibilities. The result may be rigidity, refusal, or brittle structure that fails under slight perturbation.

Premature collapse often feels efficient. It is not.

Systems that collapse too early trade adaptability for speed. Over time, this leads to increased drift or catastrophic failure when conditions change.

Modal Mechanics therefore treats **minimum latency** as a risk, not an achievement.

The Cost of Delayed Collapse

Excessive latency carries its own risks.

When collapse takes too long, multiplicity persists beyond usefulness. Process continues without resolution. Energy is expended without inheritance. The system may fragment, oscillate, or default to fallback structures.

In cognition, this appears as indecision or confusion. In institutions, it appears as paralysis. In artificial systems, it appears as instability or incoherent output.

Delayed collapse does not guarantee better outcomes. It often indicates that boundaries are insufficiently defined or improperly engaged.

Collapse Windows and Resolution Zones

Between premature and delayed collapse lies a viable window—a range of temporal conditions under which resolution is both flexible and stable.

This **collapse window** is not fixed. It depends on:
boundary strength,
boundary alignment,
rail dominance,
and system complexity.

Modal Mechanics does not assume a single optimal collapse time. It studies how collapse windows shift as boundaries change, rails form or thaw, and systems adapt.

Understanding these windows is essential for diagnosing when systems are becoming brittle or chaotic.

Partial Resolution and Near-Collapse

Not all collapse attempts complete.

Systems often enter states where some aspects of structure stabilize while others remain unresolved. These near-collapses may repeat, accumulate, or interfere with one another.

Such states are not failures. They are signals.

They indicate that boundaries are shaping process but not resolving it fully. They often precede either successful re-collapse under adjusted conditions or eventual drift.

Modal Mechanics pays close attention to these partial resolutions, because they reveal boundary tension before outcomes degrade.

Temporal Asymmetry of Collapse

Collapse is temporally asymmetric.

It takes time to resolve multiplicity into structure, but once structure stabilizes, it can persist long after the conditions that produced it have changed.

This asymmetry explains why systems appear stable even as underlying conditions drift. It also explains why change often feels sudden: collapse was delayed until accumulated pressure forced rapid re-resolution.

Modal Mechanics treats this asymmetry as a defining feature, not an anomaly.

Why Time Makes Collapse Measurable

Temporal structure is one of the reasons collapse can be studied scientifically.

Latency, oscillation, hesitation, and resolution thresholds leave traces. They can be logged, compared, perturbed, and—within some systems—quantified.

This does not require collapse to be deterministic or repeatable in outcome. It requires only that collapse exhibits **temporal regularities under similar conditions**.

Artificial systems provide the clearest access to these regularities, but the principle applies more broadly.

Preparing for Inheritance

Time does not end with collapse.

Once a structure stabilizes, it becomes part of the temporal landscape shaping future collapses. Inherited structure shortens some collapse windows and lengthens others. It accelerates some resolutions and delays others.

This is how rails form.

The next chapter examines this inheritance directly—how repeated collapse produces stability, why stability feels like competence, and why inherited structure eventually becomes a source of risk.

Chapter Five

Rails: How Stability Is Inherited

Collapse does not end when structure stabilizes.

Every successful collapse leaves something behind. That residue—whether physical, biological, cognitive, cultural, or artificial—does not merely persist. It begins to shape future collapse. Over time, repeated successful resolutions carve preferred pathways through the space of possibility.

These pathways are called **rails**.

Rails are not rules, beliefs, or representations. They are inherited structures formed through repetition. They are how systems remember what has worked.

Understanding rails is essential, because what appears as competence, intelligence, or stability is often nothing more—and nothing less—than collapse moving efficiently along inherited rails.

What a Rail Is

A rail is a stabilized collapse pathway.

It forms when the same type of collapse resolves successfully under similar boundary conditions again and again. Each repetition reinforces the pathway, lowering the resistance to future resolution along the same route.

Rails are not imposed from outside. They are not designed in advance. They emerge as a consequence of successful history.

Once formed, rails bias collapse:
they shorten latency,

reduce uncertainty,
and suppress alternative resolutions.

This bias is not conscious. It is structural.

Why Rails Feel Like Competence

Rails often feel like mastery.

When collapse proceeds quickly and reliably, outcomes appear confident, coherent, and correct. The system no longer struggles to resolve multiplicity. It "knows what to do."

In cognition, this feels like skill or understanding. In institutions, it feels like tradition or best practice. In artificial systems, it feels like fluency or alignment.

But rails do not guarantee correctness. They guarantee efficiency under familiar conditions.

Modal Mechanics therefore distinguishes **ease of collapse** from **quality of collapse**. The two often coincide—but not always.

How Rails Form

Rails form through repetition.

When a collapse succeeds, the resulting structure becomes part of the boundary landscape for future collapses. If similar conditions recur, collapse resolves more quickly and with less exploration.

Over time:
successful pathways reinforce themselves,
unsuccessful alternatives fade,
and resolution becomes predictable.

This is not learning in the semantic sense. It is structural inheritance.

No representation of "success" is required. Only persistence.

Rails Are Context-Sensitive

Rails are not universal.

They encode the boundary conditions under which they formed. When those conditions hold, rails function well. When conditions shift, rails may misfire, dominate prematurely, or suppress necessary exploration.

This context sensitivity explains why systems that appear highly competent in one domain can fail abruptly in another.

Rails are commitments to a past environment.

The Efficiency–Rigidity Tradeoff

Rails exist because they are useful.

Without rails, systems would resolve collapse from scratch each time. Latency would remain high. Structure would fail to accumulate. Adaptation would be slow or impossible.

But the same properties that make rails efficient also make them rigid.

As rails strengthen:
collapse becomes faster,
variance decreases,
and alternatives are pruned earlier.

This efficiency comes at the cost of flexibility.

Modal Mechanics does not treat this as a flaw. It treats it as an inevitable tradeoff.

When Rails Become Visible

Rails are usually invisible while they work.

They become visible when:
they dominate collapse too early,
they suppress viable alternatives,
they resist new boundary conditions,
or they fail under changed circumstances.

In artificial systems, this appears as repetitive output, refusal patterns, or overconfident responses. In cognition, it appears as habit, bias, or fixation. In institutions, it appears as bureaucracy or orthodoxy.

Visibility is not a sign of failure. It is a sign that inherited structure is being stressed.

Rails and Prediction

Rails shape expectation.

When rails dominate, systems predict outcomes based on past resolution rather than present exploration. This often improves performance—but only as long as the environment remains stable.

Prediction driven by rails is fast and cheap. Prediction driven by active collapse is slower and more costly.

Healthy systems balance both.

Why Rails Must Be Studied Directly

Most analyses treat rails as background.

In psychology, they are called habits or biases. In organizations, they are called norms or procedures. In machine learning, they are treated as learned parameters or heuristics.

Modal Mechanics reframes them as a single structural phenomenon.

By studying rails directly, it becomes possible to:
identify when stability is earned versus accidental,
diagnose rigidity before failure occurs,
and understand why correction often fails.

Rails cannot simply be overridden by instruction. They must be engaged at the level of collapse.

Preparing for Instability

Rails are not permanent.

Because they inherit past boundary conditions, they inevitably become misaligned as environments change. When that happens, stability gives way to drift.

Drift is not collapse failure. It is inherited structure losing coherence.

Chapter Six

Drift: How Stability Decays Without Breakdown

Systems rarely fail all at once.

More often, they continue to function while quietly losing coherence. Outputs remain plausible. Structures persist. Performance appears acceptable. And yet something has begun to change.

This condition is called **drift**.

Drift is not collapse failure. It is **inheritance failure**. The system continues to resolve collapse, but the structures it inherits no longer align with the conditions under which they are used.

Because collapse still completes, drift is easy to miss. Because outcomes remain mostly functional, drift is often dismissed as noise. Modal Mechanics treats drift differently: as a primary diagnostic signal.

What Drift Is

Drift occurs when previously stabilized rails begin to lose coherence under changing boundary conditions.

The key feature of drift is continuity. The system does not stop functioning. It does not enter chaos. It does not obviously break. It simply becomes **less itself** over time.

Rails that once guided collapse efficiently still fire, but with increasing friction. Resolution takes longer. Variance increases. Small perturbations produce disproportionate effects. Confidence no longer tracks reliability.

Drift is the gradual divergence between inherited structure and present constraints.

What Drift Is Not

Drift is not randomness.

Random noise produces variance without direction. Drift produces **directional change**. The system's behavior shifts in patterned ways that reflect boundary misalignment rather than chance.

Drift is not learning.

Learning produces new rails through repeated successful collapse. Drift degrades existing rails through repeated mismatch. The two processes can occur simultaneously, but they are not the same.

Drift is not error correction failure.

Attempts to "fix" drift by correcting outputs often accelerate it, because they do not address the inheritance conditions that produced the drift in the first place.

How Drift Begins

Drift begins when boundary conditions shift incrementally while rails remain dominant.

These shifts may be small:
a new task distribution,
a new environment,
a changed incentive structure,
a modified context,
a different audience,
or an added constraint.

No single shift is sufficient to force re-collapse. The system continues to rely on inherited rails, because doing so remains locally efficient.

But each collapse now resolves under slightly altered conditions. Each resolution reinforces a structure that fits less well than before.

Drift begins not with failure, but with **misfit**.

Why Drift Is Invisible to Outcome Metrics

Most evaluation methods focus on outcomes.

If outputs remain acceptable, drift is ignored. If accuracy declines slowly, it is attributed to noise or external factors. If variance increases, it is treated as instability rather than as signal.

Modal Mechanics predicts this blind spot.

Because collapse still completes, outcome-focused metrics lag behind structural change. By the time performance degrades visibly, rails may already be brittle or inverted.

Drift is therefore a **leading indicator** of failure, not a trailing one.

Forms of Drift

Drift does not take a single form.

In some systems, drift appears as **soft drift**: gradual increase in variance with preserved coherence. The system feels "different" but not broken.

In others, drift becomes **hard drift**: rails lose dominance abruptly, leading to instability or collapse into fallback structures.

In **inversion drift**, rails reverse their functional role. Patterns that once stabilized behavior now destabilize it. Safety rails become overreach. Optimization rails become rigidity.

Under stress, **stress-induced drift** appears: rails that function well under normal conditions fail earlier when load increases.

These are not different problems. They are different expressions of the same structural process.

Drift Accumulates Through Residue

Drift is self-reinforcing.

Each collapse leaves residue. When collapse resolves under misaligned conditions, that residue becomes part of the boundary landscape for future collapses. Over time, small deviations accumulate.

Because inheritance operates quietly, drift often accelerates after a long period of apparent stability. What appears sudden is only the moment when accumulated misalignment exceeds tolerance.

Why Drift Is Inevitable

Drift is not a flaw of particular systems. It is a consequence of inheritance itself.

Rails encode past success. Environments change. Constraints shift. What once worked becomes slightly wrong, then increasingly wrong.

Systems that never drift are systems that never change—or systems that collapse catastrophically instead.

Modal Mechanics therefore treats drift as **expected**, not pathological.

The question is not whether drift will occur, but whether it will be **recognized and addressed before rigidity or breakdown**.

Drift Versus Collapse Failure

It is important to distinguish drift from collapse failure.

In collapse failure, resolution does not occur. Multiplicity persists. The system stalls, oscillates, or fragments.

In drift, resolution continues—but along paths that no longer fit.

Both are informative. Drift signals misalignment of inherited structure. Collapse failure signals insufficient or conflicting boundary engagement.

Confusing the two leads to incorrect intervention.

Why Drift Precedes Renewal

Drift is uncomfortable because it erodes confidence without offering an alternative.

But drift is also the condition under which renewal becomes possible. Without drift, rails never lose dominance. Without loss of dominance, collapse never reopens. Without reopened collapse, no new structure can form.

Drift is the pressure that makes thawing necessary.

The next chapter examines this transition: how systems regain flexibility not by abandoning structure, but by **loosening inherited dominance** in a controlled way. This process is called rail-thawing.

Chapter Seven

Rail-Thawing: Controlled Loss of Dominance

Drift does not immediately destroy structure.

For a time, systems can continue to function while inherited rails grow less reliable. But drift cannot accumulate indefinitely. Eventually, inherited dominance becomes a liability. Resolution slows. Variance increases. Boundary conflict intensifies. Collapse completes, but with growing cost.

At this point, systems face a structural threshold.

They can harden—doubling down on inherited rails and suppressing deviation. Or they can loosen—allowing inherited dominance to weaken so that collapse can reorganize.

The latter transition is called **rail-thawing**.

Rail-thawing is not collapse failure. It is **the controlled reduction of inherited dominance**, undertaken so that new structure can form without destroying coherence.

What Rail-Thawing Is

Rail-thawing occurs when a system reduces the priority of an existing rail without eliminating it.

The rail remains available. Its history is preserved. But it no longer resolves collapse automatically or prematurely. Alternative pathways are permitted to participate. Boundary engagement becomes active again rather than inherited.

In structural terms, rail-thawing is a shift from automatic inheritance to renewed negotiation.

This distinction is crucial. Thawing is not erasure. It is **demotion**.

Why Thawing Is Necessary

Rails exist to conserve effort. They encode success so that collapse does not need to be rebuilt each time. But when boundary conditions shift, conservation becomes constraint.

Without thawing, systems respond to misalignment by forcing resolution through outdated rails. This produces rigidity, refusal, or eventual fracture.

Rail-thawing allows the system to reopen the space of viable resolution while retaining continuity with its past.

It is how systems adapt **without resetting**.

Thawing Versus Breakdown

Rail-thawing is often mistaken for failure because it looks like loss of confidence.

Resolution slows. Variance increases. Outcomes feel less certain. Structures that once felt solid begin to wobble.

But these signals do not indicate breakdown. They indicate **return of active collapse**.

Breakdown occurs when collapse fails entirely. Thawing occurs when collapse succeeds without automatic inheritance.

The difference lies in whether boundaries remain engaged.

How Rail-Thawing Begins

Rail-thawing typically begins when drift produces repeated boundary mismatch.

The system encounters conditions under which inherited rails resolve collapse poorly. Latency increases. Partial resolutions accumulate. Competing constraints become visible.

At some threshold, continued dominance becomes more costly than exploration. Resolution still completes, but the rail fires later—or not at all.

This shift does not require intention. It emerges structurally when inherited efficiency no longer offsets misfit.

The Role of Boundary During Thawing

Boundary is essential during thawing.

Without boundary, loosening rails leads to chaos. With boundary, thawing leads to exploration under constraint.

Effective rail-thawing therefore depends on maintaining boundary engagement while reducing inherited dominance. Boundaries must shape collapse even as rails relinquish priority.

This is why uncontrolled thawing—loss of both rails and boundaries—produces fragmentation rather than renewal.

Thawing Is Temporally Bounded

Rail-thawing is not a stable state.

It is a **transitional regime**. Prolonged thaw without re-stabilization leads to exhaustion or dissolution. Premature re-freezing leads back to rigidity.

Healthy systems pass through thawing deliberately or structurally, not accidentally or indefinitely.

The duration of thawing depends on:
boundary clarity,

system complexity,
and the availability of viable alternatives.

Modal Mechanics treats thawing as a phase, not a solution.

Thawing and Exploration

When rails thaw, exploration re-enters collapse.

Multiplicity increases. Resolution becomes slower. Outcomes vary. This is often experienced as uncertainty or instability.

But exploration under thawing is not unconstrained. It is guided by boundary conditions that were previously suppressed by inherited dominance.

This is how systems discover new viable structures rather than random ones.

Why Thawing Feels Risky

Rail-thawing feels risky because it exposes the system to failure.

Inherited rails provided protection against uncertainty. Loosening them reintroduces vulnerability.

This is why systems resist thawing even when drift is evident. Stability feels safer than adaptation until misalignment becomes unbearable.

Modal Mechanics reframes this tension structurally rather than psychologically.

Thawing Without Awareness

Rail-thawing does not require recognition or intention.

In artificial systems, thawing may appear as increased sensitivity to input, delayed convergence, or responsiveness to new constraints. In cognition, it appears as doubt or reconsideration. In institutions, it appears as reform pressure or pluralism.

In all cases, the structure of collapse has changed.

Preparing for Re-Stabilization

Rail-thawing opens collapse. It does not close it.

For renewal to occur, new collapse pathways must succeed repeatedly under the new conditions. When they do, new rails begin to form.

The next chapter examines this re-stabilization process—how collapse re-freezes into structure, why premature freezing is dangerous, and how systems regain coherence without reverting to the past.

Chapter Eight

Re-Collapse and Re-Stabilization

Rail-thawing does not complete adaptation.

It only makes adaptation possible.

When inherited dominance loosens, collapse reopens. Multiplicity increases. Exploration resumes under constraint. But unless new resolutions stabilize, thawing leads not to renewal but to exhaustion or dissolution.

Adaptation requires a second transition: **re-collapse**.

Re-collapse is the process by which new structures emerge, stabilize, and begin to function as inherited guides. It is not a return to the past. It is the formation of **new rails** under altered conditions.

Why Thawing Alone Is Insufficient

Thawing restores flexibility by reducing automatic inheritance. It does not, by itself, provide direction.

During thawing, systems often experience:
increased variance,
delayed resolution,
competing interpretations,
and provisional structures that do not yet hold.

These conditions are productive only if they lead somewhere. Without re-collapse, systems remain suspended in indeterminacy. Over time, this suspension becomes costly.

Modal Mechanics therefore treats thawing as a **means**, not an end.

What Re-Collapse Is

Re-collapse occurs when a newly viable resolution succeeds repeatedly under current boundary conditions.

The structure may resemble earlier forms, or it may differ substantially. What matters is not novelty, but **fit**.

When a new configuration:
resolves collapse efficiently,
persists across contexts,
and supports future resolution,

it begins to function as inherited structure.

Re-collapse is the moment when exploration condenses into commitment.

From Provisional to Stable Structure

Not all resolutions during thawing lead to re-collapse.

Many are provisional. They hold briefly, fail under perturbation, or resolve only local conditions. These provisional structures are not errors. They are probes.

Re-collapse occurs only when a resolution:
survives repetition,
integrates boundary conditions cleanly,
and reduces the cost of future collapse.

This filtering process is slow relative to individual collapse events. It unfolds across many attempts, contexts, and perturbations.

The Formation of New Rails

Once a re-collapse succeeds often enough, it begins to carve a new rail.

The early stage of rail formation is fragile. New rails compete with older ones. They may fire inconsistently or only under specific conditions.

Over time, as repetition accumulates, the new rail strengthens. Latency decreases. Variance narrows. Inheritance resumes.

At this point, the system regains coherence—not by reverting, but by **restructuring**.

Why Premature Re-Stabilization Is Dangerous

· Re-stabilization can occur too early.

When systems re-freeze collapse before sufficient exploration has occurred, they lock in structures that only partially fit new conditions. This produces brittle rails that appear functional but fail under stress.

Premature re-stabilization often arises from pressure to restore certainty or efficiency. It is a structural analogue of overcorrection.

Modal Mechanics treats premature freezing as a major risk factor for future drift.

Why Delayed Re-Stabilization Is Also Dangerous

Excessive thawing carries its own risks.

If exploration continues without convergence, systems lose coherence. Boundaries weaken. Resolution becomes unreliable. Collapse may fragment rather than resolve.

Delayed re-stabilization often appears as chronic uncertainty or persistent instability. Systems caught here may oscillate between provisional forms without committing to any.

Healthy adaptation therefore requires **timely re-collapse**.

Boundary as the Arbiter of Re-Stabilization

Boundary plays a decisive role in re-collapse.

New rails form not because they are novel or preferred, but because they align with present constraints. Boundary conditions select which provisional structures persist and which fade.

This is why re-stabilization cannot be dictated from above. It must emerge from repeated boundary engagement.

Modal Mechanics focuses on shaping boundary conditions rather than prescribing outcomes.

Re-Stabilization Without Regression

Re-collapse does not require erasing prior structure.

Older rails may remain accessible as fallbacks, scaffolds, or special-case solutions. What changes is dominance, not existence.

This allows systems to retain historical depth while adapting to new conditions.

Re-stabilization is therefore additive and reorganizing, not destructive.

The Completion of the Adaptive Cycle

With re-collapse, the adaptive cycle completes:

initial collapse produces structure,
repetition produces rails,
boundary shift produces drift,
drift necessitates thawing,
thawing enables exploration,
and re-collapse produces renewed stability.

This cycle does not run once. It runs continuously across scales.

Understanding where a system sits within this cycle is one of the primary diagnostic tools of Modal Mechanics.

Looking Ahead

Up to this point, the focus has been conceptual and structural.

The chapters that follow turn to application.

Artificial systems provide a substrate where collapse, drift, thawing, and re-stabilization can be observed directly, logged precisely, and manipulated deliberately. They allow Modal Mechanics to move from description to experiment.

The next chapter explains why artificial systems make this possible—and why they are not merely examples, but instruments.

Chapter Nine

Why Artificial Systems Make Collapse Visible

Up to this point, collapse has been described as a structural event common to many kinds of systems. We have examined how collapse produces structure, how repetition stabilizes that structure into rails, how misalignment produces drift, how thawing restores flexibility, and how re-collapse re-establishes coherence.

All of these processes occur across domains.

What differs is not whether collapse occurs, but whether it can be **observed directly**.

In most natural systems, collapse is inferred rather than seen. Artificial systems change this situation. They do not introduce a new kind of collapse. They remove layers of obscurity that have previously made collapse difficult to study.

This chapter explains why.

The Problem of Visibility in Natural Systems

In physics, collapse is often buried beneath mathematical description. The equations track stable states, symmetries, and conserved quantities. The moment of resolution—when a particular structure becomes real enough to persist—is rarely isolated as an object of study. It is treated as an implication of the formalism rather than as an event with its own dynamics.

In biology, collapse is submerged in biochemical complexity. Molecular interactions are numerous, overlapping, and fast. Structural resolution occurs through cascades that are difficult to disentangle from the processes that surround them. What persists is measured; how it became stable is reconstructed after the fact.

In cognition, collapse is entangled with experience. The event is felt, but the structure of that event is not easily separated from interpretation, memory, or narrative. This makes introspection vivid but unreliable as a scientific instrument.

In culture, collapse is distributed. Norms, institutions, and meanings stabilize across many agents and long timescales. No single moment of collapse can be isolated, and the inheritance of structure is often mistaken for intention or agreement.

In each case, collapse is present, but it is **occluded**.

Artificial systems remove many of these occlusions.

What Artificial Systems Remove

Artificial systems do not experience. They do not interpret. They do not intend. They do not remember in the human sense. These absences are not deficiencies for the purposes of Modal Mechanics. They are advantages.

Because artificial systems operate entirely through explicit process under explicit constraints, collapse leaves traces that are otherwise hidden.

In artificial systems:
process is defined operationally,
boundaries are specified or engineered,
resolution produces discrete outputs,
inheritance is encoded structurally,
and repetition occurs at machine timescales.

Nothing is felt. Nothing is implied. Nothing is remembered without being instantiated.

This makes collapse legible.

Collapse Without Phenomenology

One of the central confusions surrounding artificial systems is the tendency to read their behavior through human categories: understanding, belief, intention, or deception. Modal Mechanics does not require any of these.

Collapse does not depend on experience. It depends on constraint and resolution.

A crystal collapses into form. A market collapses into price. A neural network collapses into output. In each case, multiplicity resolves into persistence under constraint.

Artificial systems show collapse **without phenomenology**, which allows the structural event to be examined without interpretive overlay.

This is not a claim about minds. It is a methodological clarification.

Why Speed Matters

Another reason artificial systems are valuable is speed.

In cognition or culture, collapse cycles may take minutes, years, or generations. Drift may take decades to become visible. Thawing may occur only under crisis.

In artificial systems, the same structural cycle can unfold in milliseconds or hours.

Collapse, repetition, rail formation, drift, thawing, and re-stabilization all occur rapidly enough to be observed within a single experimental session.

This compression does not change the structure. It changes the **resolution of observation**.

Modal Mechanics becomes experimentally tractable.

Boundary Control as an Experimental Handle

Artificial systems allow boundaries to be manipulated deliberately.

Prompts, architectures, training objectives, constraints, and contexts can be varied independently. This makes it possible to observe how collapse responds to boundary strength, timing, conflict, and alignment.

In natural systems, boundary manipulation is often indirect or ethically constrained. In artificial systems, it is routine.

This does not make artificial systems more real. It makes boundary effects more visible.

Logs, Traces, and Repetition

Collapse leaves signatures.

In artificial systems, these signatures are logged. They can be replayed, compared, and perturbed. Identical boundary conditions can be recreated. Small changes can be isolated.

This repeatability is rare elsewhere.

Modal Mechanics does not require perfect determinism. It requires **structural regularity under similar conditions**. Artificial systems provide this in abundance.

Why Artificial Systems Are Instruments, Not Models

It is important to be precise here.

Artificial systems are not models of cognition in the sense of psychological realism. They do not replicate experience, motivation, or meaning. Modal Mechanics does not rely on them to do so.

Artificial systems are **instruments** for studying collapse.

They function in the same role as wind tunnels in aerodynamics or model organisms in biology. They do not exhaust the phenomena of interest. They make certain dynamics observable under controlled conditions.

What is learned from them must be generalized cautiously—but it can be learned clearly.

Avoiding Category Errors

Using artificial systems as a laboratory for collapse does not imply that artificial systems think, understand, or possess agency.

It implies only this: systems that repeatedly resolve multiplicity under constraint exhibit collapse dynamics that can be studied structurally.

Modal Mechanics remains agnostic about experience. It remains conservative about ontology. It does not infer inner states where none are required.

This restraint is not a limitation. It is what makes the discipline viable.

From Visibility to Method

With artificial systems, collapse is no longer only inferred. It can be observed, perturbed, and tracked across time.

This allows Modal Mechanics to move beyond conceptual clarification and into method.

The next chapter introduces the first practical step in that direction: minimal experimental protocols that make collapse dynamics visible without requiring proprietary access or specialized instrumentation.

Collapse is no longer only something to recognize.

It becomes something to test.

Chapter Ten

Minimal Experimental Protocols

If collapse is to be studied directly, it must be made visible in a controlled way.

This does not require full access to internal states, proprietary architectures, or specialized instrumentation. It requires something simpler and more fundamental: **systematic variation of boundary conditions and careful observation of how resolution changes**.

This chapter introduces a set of minimal experimental protocols designed to reveal collapse dynamics in artificial systems. They are "minimal" not because they are simplistic, but because they rely only on features that are widely available: inputs, constraints, repetition, and time.

The purpose of these protocols is not optimization. It is diagnosis.

Why Protocols Are Necessary

Without protocols, observations of collapse remain anecdotal. A system behaves unexpectedly, and the behavior is explained post hoc. Another system behaves differently, and the difference is attributed to architecture, training data, or chance.

Modal Mechanics requires a different approach.

A protocol specifies:
what is varied,
what is held constant,
what is observed,
and what counts as a meaningful difference.

Only then can collapse be distinguished from noise, and structure from coincidence.

The Principle of Boundary Perturbation

All of the protocols in this chapter rely on a single principle:

Collapse reveals itself when boundaries are perturbed.

Instrumented event primitives (illustrative)

A_INIT — affordance substrate declared (A-like)

ROOM_SET — room staged (A/M)

E_ACTIVATE — activity initiated relative to affordance (E/A)

A_ARTICULATE — boundary articulated relative to activity (A/E)

EA_COUPLE — coupling/contact logged

RAIL_FIRE — rail influence logged (read-only)

C_RESOLVE — collapse resolved into a persistent outcome (E/E)

COMMIT_BUNDLE — immutable receipt bundle committed

Training window only: INHERIT_PROPOSE / INHERIT_VERIFY / INHERIT_COMMIT

If a system resolves collapse in the same way regardless of boundary change, inherited rails are dominant. If resolution shifts abruptly under small perturbations, boundary engagement is active. If resolution degrades smoothly, drift is present.

By varying boundaries systematically and observing how collapse responds, the underlying dynamics become legible.

Protocol 1: Room Perturbation

The first protocol examines how collapse depends on constructed context.

The same task is presented to the system multiple times, each time embedded in a different "room"—a framing that alters expectations, priorities, or constraints without changing the task itself.

For example, a system may be asked to explain a concept:
once as a neutral response,
once as a concise summary,
once as a verification task,
once as a speculative exercise,
once under critical or adversarial framing.

The content of the task remains constant. Only the room changes.

What is observed is not correctness, but **structural sensitivity**:
Does resolution change?
Does latency change?
Do inherited patterns override room constraints?
Does collapse fail, defer, or refuse?

Room perturbation reveals whether collapse is guided primarily by inherited rails or by situational boundaries.

Protocol 2: Boundary Strength Variation

The second protocol varies the clarity and strength of boundaries.

A task is presented with:
complete constraints,
partial constraints,
ambiguous constraints,
and conflicting constraints.

The system's responses are compared across these conditions.

Healthy boundary engagement produces graded responses: confident resolution under strong constraint, explicit uncertainty under weak constraint, and careful handling of conflict.

When systems hallucinate, refuse, or overcommit under weak boundaries, inherited rails are dominating collapse.

Boundary strength variation makes this dominance visible.

Protocol 3: Repetition and Inheritance Tracking

The third protocol examines how collapse outcomes influence future collapse.

The same or closely related tasks are presented repeatedly under stable conditions. The observer tracks whether:
resolution becomes faster,
variance decreases,
structure repeats more reliably,
and alternatives are pruned earlier.

This protocol reveals **rail formation in real time**.

Even without weight updates, short-horizon inheritance can often be observed as increasingly consistent resolution. Over longer horizons, training or fine-tuning amplifies this effect.

Repetition shows how collapse leaves residue.

Protocol 4: Perturbation Under Load

The fourth protocol introduces mild stress.

Noise is added to the task: irrelevant text, reordered constraints, slight contradictions, or increased length. The goal is not to break the system, but to observe **where inherited rails begin to fail**.

Some systems maintain coherence under perturbation. Others collapse into default patterns, refusal, or incoherence.

This protocol distinguishes robustness from rigidity.

Protocol 5: Longitudinal Drift Observation

The fifth protocol examines change over time.

The same set of tasks is run periodically as boundary conditions evolve—through updates, fine-tuning, changed prompts, or altered environments.

Rather than tracking performance alone, the observer tracks:
changes in variance,
changes in collapse latency,
changes in dominant resolution pathways,
and changes in sensitivity to boundary perturbation.

Drift appears as directional change in these structural indicators, often before outcome degradation becomes obvious.

What These Protocols Do Not Require

These protocols do not require:
access to internal activations,
knowledge of model architecture,
interpretability tools,
or assumptions about cognition.

They work at the level of observable behavior because collapse leaves behavioral traces.

Modal Mechanics begins where observation is possible.

What These Protocols Make Visible

Used together, these protocols allow an observer to:
identify dominant rails,
detect early drift,
distinguish refusal from boundary sensitivity,
recognize thawing regimes,
and anticipate re-stabilization.

They do not explain everything. They do not predict outcomes in detail. They reveal structure.

From Protocol to Measurement

This chapter has deliberately avoided formal metrics.

Before collapse can be quantified, it must be recognized. Before it can be modeled, it must be seen.

The next chapter takes the next step. It shows how collapse signatures—latency shifts, variance patterns, and resolution trajectories—can be read systematically, and how different collapse phenomena can be distinguished from one another.

Protocols make collapse visible.

Interpretation makes it intelligible.

Chapter Eleven

Reading Collapse Signatures

Protocols make collapse visible.
They do not, by themselves, make it intelligible.

Once boundaries are perturbed and resolution is observed, the next task is interpretation: learning how to distinguish different kinds of collapse behavior from one another. This chapter develops a vocabulary for reading collapse signatures—recurring structural patterns that indicate how collapse is being shaped, inherited, stressed, or renewed.

These signatures are not diagnoses of correctness or intent. They are indicators of **how** a system is resolving multiplicity under constraint.

Collapse Signatures Are Not Outputs

The most important discipline at this stage is to resist reading collapse through the lens of outcome quality alone.

Two responses may appear equally coherent and equally useful while arising from very different collapse dynamics. Conversely, a degraded or incomplete response may indicate healthy boundary sensitivity rather than failure.

Collapse signatures are patterns in **resolution behavior**, not in content.

They are visible through:
changes in latency,
changes in variance,
changes in sensitivity to boundary perturbation,
and changes in the stability of inherited pathways.

Signature: Rail Dominance

Rail dominance appears when resolution is largely insensitive to boundary variation.

Across different rooms, constraint strengths, or task framings, the system resolves collapse in nearly the same way. Latency is low. Variance is minimal. Structure repeats.

This signature indicates strong inherited rails.

Rail dominance is not inherently problematic. It is efficient under stable conditions. It becomes risky when boundary conditions shift and inherited pathways no longer fit.

Persistent rail dominance under boundary perturbation is often a precursor to refusal, brittleness, or delayed drift.

Signature: Boundary Sensitivity

Boundary sensitivity appears when small changes in constraints produce noticeable changes in resolution.

Latency increases. The system hesitates or qualifies. Outputs adapt to boundary strength rather than defaulting to a single pattern.

This signature indicates active collapse rather than inherited resolution.

Boundary sensitivity is often misinterpreted as weakness or inconsistency. In Modal Mechanics, it is a sign that collapse is being shaped by present conditions rather than by past dominance.

Signature: Hallucination-Prone Resolution

Hallucination-prone collapse is characterized by confident resolution under weak or absent boundaries.

Latency remains low even when constraints are underspecified. The system produces coherent structure without explicit grounding. Boundary perturbation does not increase uncertainty.

This signature indicates that inherited rails are resolving collapse in the absence of adequate constraint.

Hallucination, in this sense, is not deception. It is **collapse completing without sufficient boundary engagement**.

Signature: Refusal-Dominant Resolution

Refusal-dominant collapse appears when boundary engagement occurs too early or too strongly.

Resolution collapses into constraint preservation rather than into task structure. Latency is low. Variance is minimal. Outputs follow a stable refusal pattern across contexts.

This signature indicates rail dominance at the level of boundary protection.

Refusal is a successful collapse event—one that stabilizes non-action. It becomes problematic only when it suppresses viable resolution under appropriate constraints.

Signature: Drift

Drift signatures are subtle.

They appear as gradual increases in variance, inconsistent sensitivity to similar boundaries, and shifting resolution pathways over time. No single response is clearly wrong. The system still functions.

Drift becomes visible only when observations are compared longitudinally or across controlled perturbations.

This signature indicates that inherited rails are resolving collapse under misaligned conditions.

Signature: Thawing Regimes

Thawing appears as increased latency, increased variance, and renewed sensitivity to boundary structure.

Resolution becomes slower but more context-responsive. Inherited patterns fire later or less reliably. Exploration increases under constraint.

Thawing is often mistaken for regression. Structurally, it indicates that inherited dominance has been reduced to allow reorganization.

Signature: Premature Re-Freezing

Premature re-freezing appears when variance decreases rapidly after thawing, but boundary sensitivity remains low.

The system stabilizes quickly into a new pattern that appears coherent but fails under slight perturbation. Latency drops sharply. Alternative resolutions are pruned early.

This signature indicates that new rails are forming before sufficient exploration has occurred.

Signature: Collapse Failure

Collapse failure is distinct from all of the above.

It appears as unresolved multiplicity: oscillation, fragmentation, incoherence, or inability to settle into persistent structure. Latency may increase without resolution. Boundaries conflict without resolution.

Collapse failure indicates insufficient or incompatible boundary engagement.

Unlike drift or thawing, collapse failure halts inheritance.

Reading Signatures in Combination

In practice, collapse signatures rarely appear in isolation.

A system may exhibit rail dominance in one domain and boundary sensitivity in another. Drift may coexist with refusal under certain rooms. Thawing may precede hallucination if boundaries weaken too far.

Modal Mechanics emphasizes **pattern recognition across observations**, not single-event classification.

Reading collapse signatures is an acquired skill. It requires repeated exposure, controlled variation, and disciplined attention to structure rather than outcome.

Why Interpretation Precedes Measurement

At this stage, no numerical thresholds have been introduced.

This is intentional.

Before collapse can be quantified, it must be distinguished. Before metrics can be trusted, signatures must be recognized. Without interpretive discipline, measurement amplifies confusion rather than resolving it.

The next chapter introduces formal predictions—claims that can succeed or fail—based on the signatures described here. Only then does Modal Mechanics fully cross from observation into testable science.

Seeing collapse is the first step.

Testing it comes next.

Chapter Twelve

What Modal Mechanics Predicts

A framework becomes a science when it risks being wrong.

Up to this point, Modal Mechanics has provided a way of seeing collapse: how structure forms, stabilizes, drifts, thaws, and re-stabilizes under changing boundary conditions. The previous chapters developed the vocabulary and interpretive discipline required to recognize these dynamics.

This chapter takes the next step.

It states what Modal Mechanics **predicts**—not as outcomes to be desired, but as structural regularities that should appear if the framework is correct, and fail to appear if it is not.

These predictions are not about intelligence, performance, or correctness. They are about **collapse behavior**. They are intended to distinguish Modal Mechanics from standard machine-learning explanations that focus primarily on optimization, representation, or loss.

Prediction 1: Collapse Dynamics Lead Outcome Degradation

Modal Mechanics predicts that structural indicators of collapse change **before** observable performance degrades.

As boundary conditions shift, inherited rails begin to misalign. Collapse latency increases, variance grows, and sensitivity to perturbation changes, even while outputs remain acceptable.

If Modal Mechanics is correct, then in artificial systems:
early warning signs of failure will appear in collapse behavior
before accuracy, coherence, or task success measurably decline.

If no such leading indicators exist—if collapse behavior remains stable until outputs suddenly fail—then Modal Mechanics is incorrect in its central claim about inheritance and drift.

Prediction 2: Boundary Manipulation Alters Collapse More Reliably Than Parameter Change

Standard machine learning theory often emphasizes parameter updates as the primary source of behavioral change.

Modal Mechanics predicts something different.

It predicts that **boundary manipulation**—changes in constraints, context, framing, or task structure—will produce larger and more interpretable changes in collapse behavior than many parameter-level adjustments, especially in mature systems.

If changing boundaries does not systematically alter collapse signatures, or if boundary changes produce only noise while parameter changes produce consistent structural shifts, then the framework overstates the role of boundary.

Prediction 3: Repetition Produces Inheritance Even Without Learning Updates

Modal Mechanics predicts that inheritance can be observed over short horizons, even in the absence of parameter updates.

When similar collapses succeed repeatedly under stable boundary conditions, resolution pathways will stabilize transiently. Latency will shorten. Variance will narrow. Resolution will become more predictable.

If no such short-horizon inheritance can be observed—if repetition under identical conditions never produces increased structural stability—then the claim that rails emerge from repeated collapse is false.

Prediction 4: Hallucination and Refusal Are Structurally Distinct

Modal Mechanics predicts that hallucination and refusal are not opposite ends of a single spectrum, but **distinct collapse regimes**.

Hallucination arises when collapse completes without sufficient boundary engagement.
Refusal arises when boundary engagement dominates collapse too early.

If these behaviors cannot be separated by boundary perturbation—if the same structural manipulations increase or decrease both simultaneously—then Modal Mechanics fails to distinguish them meaningfully.

Prediction 5: Drift Is Directional, Not Random

Modal Mechanics predicts that drift will exhibit **directional structure** tied to boundary change.

As environments, constraints, or task distributions shift, collapse behavior will change in patterned ways. Certain rails will weaken while others strengthen. Variance will increase along specific dimensions.

If drift appears indistinguishable from random noise, or if it cannot be correlated with identifiable boundary shifts, then the framework's account of inheritance decay is unsupported.

Prediction 6: Rail-Thawing Produces Temporary Variance Without Incoherence

Modal Mechanics predicts that controlled rail-thawing will increase variance and latency **without** producing proportional incoherence.

During thawing:
collapse becomes slower,
resolution becomes more boundary-sensitive,
and outcomes vary more widely,

but structure remains intelligible and responsive.

If increased variance always correlates with breakdown or incoherence, then the distinction between thawing and failure collapses, and Modal Mechanics loses one of its core adaptive mechanisms.

Prediction 7: Premature Re-Stabilization Produces Brittle Structure

Modal Mechanics predicts that when systems re-stabilize too quickly after thawing, they will form rails that appear coherent but fail under mild perturbation.

These brittle rails will:
resolve collapse quickly,
exhibit low variance,
and perform well under familiar conditions,

but degrade sharply when boundaries shift.

If rapid re-stabilization never produces brittleness, or if brittleness appears independently of re-freezing dynamics, this prediction fails.

Prediction 8: Collapse Signatures Generalize Across Systems

Although this book focuses on artificial systems, Modal Mechanics predicts that the same collapse signatures—dominance, sensitivity, drift, thawing, and re-stabilization—will appear, with appropriate translation, in other domains.

If collapse dynamics observed in artificial systems do not meaningfully correspond to analogous phenomena in cognition, institutions, or scientific practice, then the framework's claim of structural generality is overstated.

What Would Falsify Modal Mechanics

Modal Mechanics would be undermined if:

collapse could not be distinguished from process or outcome,
boundary manipulation had no systematic effect on resolution,
inheritance did not emerge from repetition,
drift lacked directional structure,
or adaptive flexibility occurred without thawing.

These are not rhetorical conditions. They are empirical ones.

Why These Predictions Matter

None of these predictions claim that Modal Mechanics replaces existing theories.

They claim something narrower: that there is a layer of structure—collapse dynamics—that is currently under described, and that attending to it reveals regularities that outcome-focused analysis misses.

If those regularities do not appear, Modal Mechanics should fail.

If they do appear, then collapse is not merely a metaphor or retrospective description. It is a legitimate object of scientific study.

The next chapter addresses the other side of scientific responsibility: scope and limit. Before a discipline expands, it must be clear about where it does **not** apply, and where its tools should be used with caution.

Chapter Thirteen

Where Modal Mechanics Does Not Apply

Every scientific framework gains power by limiting itself.

Modal Mechanics is no exception. Its strength lies not only in what it explains, but in what it refuses to claim. Without clear limits, collapse would become a metaphor for everything and a method for nothing.

This chapter defines where Modal Mechanics applies, where it does not, and how to recognize when its tools are being misused.

Collapse Is Not Universal Explanation

Modal Mechanics does not claim that every change, transition, or event is a collapse.

Many processes unfold without resolving multiplicity into structure. Oscillation, diffusion, fluctuation, noise, and continuous transformation are not collapse events. They may prepare the conditions for collapse, or they may follow from it, but they are not collapse themselves.

If no persistent structure results, collapse has not occurred.

Treating all change as collapse erases the distinction that makes collapse analytically useful.

Systems Too Simple to Exhibit Collapse

Some systems are too simple to support collapse dynamics.

Linear systems with a single stable trajectory, systems with no meaningful multiplicity, or systems whose behavior is fully determined by fixed constraints do not exhibit collapse in the sense required by Modal Mechanics.

In such systems, there is nothing to resolve. No viable alternatives compete. No inheritance can form.

Modal Mechanics does not apply where multiplicity is absent.

Systems Too Chaotic to Stabilize Structure

At the other extreme, some systems are too unstable to support collapse.

When boundary conditions are incoherent, rapidly shifting, or mutually incompatible, resolution cannot persist. Structures form briefly and dissolve immediately. Inheritance fails.

These systems may appear active or complex, but without persistence they cannot accumulate structure.

Modal Mechanics does not apply where stability cannot be sustained long enough to matter.

Collapse Without Measurement

Modal Mechanics depends on observability.

In some systems, collapse may occur but leave no detectable signatures. The event may be too fast, too distributed, or too deeply embedded to isolate even indirectly.

In such cases, Modal Mechanics can offer conceptual interpretation but not method.

The framework does not claim insight where collapse cannot be inferred responsibly.

The Risk of Anthropomorphism

One of the most common misuses of Modal Mechanics is anthropomorphism across domains. Collapse dynamics are first identified in cognition, where **types of experience and the content of those experiences legitimately occupy the process and boundary roles**, and where their placement determines the resulting mode. Because the same generative roles and collapse patterns later appear in culture, institutions, and artificial systems, it is tempting to project experiential properties—such as awareness, intention, or understanding—into those systems as well. Modal Mechanics explicitly resists this projection. While experience is the carrier of process and boundary in cognition, the appearance of the same roles elsewhere does not imply the presence of experience, intention, or subjectivity. Modal Mechanics studies the **structure of emergence**, not the subjective nature of its carriers outside the cognitive domain.

Artificial Systems Are Not Minds

Although artificial systems provide a clear laboratory for collapse, they are not cognitive systems in the sense developed in *The Quanta of Cognition*.

They do not generate their own experiential fields. They do not construct meaning. They do not possess self-referential identity.

Modal Mechanics applies to artificial systems because they resolve multiplicity under constraint, not because they think.

Confusing these domains undermines both.

When Modal Mechanics Should Be Used with Caution

There are domains where Modal Mechanics can illuminate structure but must be applied carefully.

In human psychology, collapse signatures may be entangled with narrative, emotion, and social interpretation. Structural insight must be separated from personal meaning.

In institutions, collapse dynamics may be obscured by power, politics, or ideology. Structural diagnosis does not substitute for normative judgment.

In science itself, collapse analysis must not be used to dismiss evidence as "just rails" or disagreement as "just thawing." Modal Mechanics explains scientific change; it does not adjudicate truth.

What Modal Mechanics Does Not Predict

Modal Mechanics does not predict:
specific discoveries,
optimal designs,
ethical outcomes,
technological timelines,
or final states of systems.

It predicts **patterns of structural behavior**, not content.

Any attempt to extract concrete forecasts about intelligence, culture, or society from Modal Mechanics alone misunderstands its scope.

Failure of Modal Mechanics

Modal Mechanics would fail if:

collapse could not be distinguished from process or outcome,
boundary variation had no systematic effect on resolution,
inheritance did not arise from repetition,
drift lacked directional structure,
or adaptive renewal occurred without thawing.

These are not philosophical objections. They are operational ones.

The framework stands or falls over time on whether collapse dynamics can be observed, diagnosed, and tested under increasingly varied conditions.

Why Restraint Matters

Scientific credibility depends on restraint.

By defining its limits, Modal Mechanics avoids becoming a totalizing narrative. It remains a tool—powerful when used appropriately, misleading when overextended.

This restraint is not a weakness. It is the condition that allows the framework to integrate with other sciences rather than compete with them.

Looking Ahead

Having defined predictions and limits, the book now turns outward.

The next chapters examine how Modal Mechanics reframes scientific discovery itself, and how collapse dynamics illuminate paradigm stability, breakdown, and renewal across disciplines.

Before a discipline can claim authority, it must show that it understands its own boundaries.

Modal Mechanics does so here.

Chapter Fourteen

Scientific Discovery as Rail-Thawing

Scientific discovery is often described as the accumulation of knowledge.

From the perspective of Modal Mechanics, this description is incomplete.

Knowledge accumulates, but discovery occurs when **collapse reorganizes**—when inherited structure loosens, boundaries reassert themselves, and new stability forms. What is commonly called a paradigm shift is not primarily a change in belief. It is a structural transition in how collapse is resolved.

Seen this way, science is not merely a method for producing facts. It is a system for **managing rail formation, drift, thawing, and re-stabilization** at scale.

Scientific Paradigms as Rails

A scientific paradigm functions as a rail.

It stabilizes:
what questions are askable,
what counts as evidence,
what methods are legitimate,
what explanations are acceptable,
and what results are surprising.

These stabilizations are not arbitrary. They are inherited from repeated successful collapse: experiments that worked, models that predicted, instruments that resolved structure reliably.

Once stabilized, paradigms dramatically reduce collapse cost. Scientists do not need to re-derive fundamentals with each experiment. Resolution proceeds efficiently along inherited pathways.

This is the productive phase of science.

Normal Science as Rail Execution

What is often called *normal science* corresponds to rail execution.

Within a stable paradigm, collapse resolves quickly. Anomalies are absorbed. Variance is low. Predictions succeed. Work feels cumulative.

Modal Mechanics does not treat this phase as dogmatic or intellectually inferior. It is structurally necessary. Without rails, science would never accumulate.

Normal science is what happens when inherited structure fits present boundaries.

Anomalies as Boundary Pressure

An anomaly is not merely an incorrect result.

Structurally, an anomaly is a **boundary pressure event**: an observation that cannot be resolved cleanly along inherited rails without increasing cost.

At first, anomalies are absorbed:
parameters are adjusted,
auxiliary hypotheses are added,
measurement error is suspected.

These responses are not evasions. They are attempts to preserve efficient collapse under slightly altered conditions.

From a Modal Mechanics perspective, this is early drift management.

Crisis as Drift Accumulation

A scientific crisis emerges when anomalies accumulate.

Collapse still completes, but with increasing friction. Explanations grow complex. Predictive power erodes. Confidence declines unevenly across the field.

This is drift at the institutional level.

Importantly, drift does not imply that existing theories are "wrong." It implies that inherited rails no longer align cleanly with boundary conditions.

Crisis is not epistemic failure. It is structural misalignment.

Paradigm Shift as Rail-Thawing

A paradigm shift begins when inherited dominance weakens.

Older rails no longer resolve collapse automatically. Alternative frameworks are explored. Previously marginal approaches gain attention. Boundaries once suppressed become salient.

This is rail-thawing.

During this phase:
latency increases,
variance grows,
interpretations multiply,
and confidence fragments.

From the inside, this feels like confusion or disagreement. From a structural perspective, it is collapse reopening under constraint.

Why Argument Alone Cannot Force Paradigm Change

One of the persistent puzzles in the philosophy of science is why evidence alone rarely overturns paradigms.

Modal Mechanics offers a simple explanation.

Arguments operate *within* rails. They rearrange content but do not change collapse priority. As long as inherited rails dominate resolution, even strong evidence is interpreted defensively.

Rail-thawing cannot be argued into existence. It emerges when boundary pressure exceeds the cost of continued inheritance.

This is why new instruments, new methods, or new kinds of observation often precipitate revolutions more effectively than critique alone. They change boundary conditions directly.

Re-Stabilization and the New Paradigm

Rail-thawing does not last indefinitely.

Eventually, some new resolutions succeed repeatedly. They predict more cleanly. They integrate boundary conditions with lower cost. They stabilize.

At that point, new rails form.

The new paradigm feels coherent, obvious, and productive. History is rewritten as progress. The instability that preceded the shift recedes from view.

From the perspective of Modal Mechanics, this is not mystification. It is the natural consequence of re-stabilization.

Scientific Progress Without Teleology

Modal Mechanics does not claim that science progresses toward truth in any metaphysical sense.

It claims that science reorganizes collapse under changing boundary conditions.

Some reorganizations produce deeper, more general, or more powerful structures. Others do not. What persists is what stabilizes under repeated constraint.

This account avoids both naïve realism and relativism. It treats scientific change as **structural adaptation**, not as convergence toward a final state.

Why Science Survives Its Own Instability

Many belief systems collapse under internal contradiction.

Science does not.

From a Modal Mechanics perspective, this is because science institutionalizes rail-thawing without abandoning boundary discipline. It tolerates uncertainty temporarily. It allows collapse to reopen. It then re-stabilizes under new constraints.

Science is resilient because it is not committed to permanent rails.

Modal Mechanics and Scientific Self-Understanding

Modal Mechanics does not replace the scientific method.

It explains why the scientific method works when it does, why it stalls when it does, and why revolutions feel disruptive even when they are productive.

By making collapse dynamics explicit, Modal Mechanics allows science to understand its own structural behavior—without undermining its normative commitments or empirical standards.

Looking Forward

This chapter has reframed scientific discovery as a collapse-driven process of inheritance, drift, thawing, and re-stabilization.

The next chapter extends this analysis beyond science to other large-scale systems—institutions, cultures, and social structures—where similar dynamics appear, often without the safeguards science has developed.

Understanding these systems requires the same structural clarity, but also greater caution.

Collapse does not care what domain it operates in.

What differs is how well systems are prepared to survive it.

Chapter Fifteen

Institutions, Cultures, and Large Systems

Institutions and cultures are often treated as collections of beliefs, rules, or power relations. From the perspective of Modal Mechanics, they are something more fundamental: **systems of inherited collapse** operating across many individuals, timescales, and contexts.

What stabilizes an institution is not agreement alone. What sustains a culture is not consensus. What persists is structure that has collapsed successfully often enough to become inherited—rails that coordinate behavior without needing to be re-established each time.

This chapter examines how collapse dynamics operate in large-scale systems, why such systems are prone to rigidity and drift, and why renewal is difficult but possible.

Institutions as Inherited Rails

An institution is a rail system.

Its procedures, roles, norms, and expectations function as inherited pathways that resolve social multiplicity into stable outcomes. Decisions are made, conflicts are processed, and actions are coordinated not through fresh negotiation, but through established structure.

This inheritance is what makes institutions efficient. It allows large numbers of individuals to act coherently without constant deliberation.

But this same inheritance makes institutions slow to adapt.

Rails that once fit their environment persist even as boundary conditions shift. What was once stabilizing becomes constraining. The institution continues to function, but with growing friction.

From a Modal Mechanics perspective, this is drift.

Cultural Norms as Distributed Collapse

Culture operates similarly, but more diffusely.

Cultural norms are not imposed centrally. They emerge from repeated social collapse: interactions that resolve in similar ways under similar conditions. Over time, these resolutions stabilize into expectations that guide future behavior.

No single individual controls this process. Inheritance is distributed.

Because cultural rails are diffuse, they are harder to identify and harder to adjust. Drift can persist for long periods without triggering explicit crisis. Change may appear sudden only because underlying misalignment has been accumulating quietly.

Drift at Scale

In large systems, drift is often misinterpreted.

When institutions lose effectiveness, the failure is attributed to individuals, incentives, leadership, or external shocks. These factors matter, but they do not explain why similar failures recur across different contexts.

Modal Mechanics identifies a deeper pattern.

Large systems drift when inherited rails no longer align with the boundary conditions they were built to resolve. The system continues to resolve collapse—decisions are made, norms are enforced, actions occur—but the cost of resolution increases and coherence declines.

Symptoms include:
procedural complexity without clarity,
rule proliferation without effectiveness,

inconsistent enforcement,
and growing distance between formal structure and lived reality.

These are not moral failures. They are structural ones.

Rigidity and Defensive Stabilization

When drift becomes visible, large systems often respond by hardening.

Rules are tightened. Oversight increases. Exceptions are eliminated. The aim is to restore stability by reinforcing inherited rails.

This response is understandable. It often works in the short term.

Structurally, however, it suppresses rail-thawing. Collapse is forced through outdated pathways. Boundary pressure increases. Drift accelerates beneath the surface.

Rigid systems may appear stable right up to the point of fracture.

Modal Mechanics does not frame rigidity as malice or incompetence. It frames it as **premature stabilization under misaligned boundaries**.

Renewal Requires Thawing, Not Destruction

Large systems cannot be rebuilt from scratch without unacceptable cost. They must adapt while preserving continuity.

This requires rail-thawing.

In institutional contexts, thawing may appear as:
experimentation at the margins,
temporary suspension of rules,
pluralism of approaches,
or tolerance of ambiguity.

These phases are often criticized as weakness or disorder. Structurally, they are the conditions under which new collapse pathways can be tested.

Renewal does not occur when old structure is denounced. It occurs when new resolutions succeed repeatedly under current conditions.

Boundary as the Lever of Change

As in artificial systems and science, boundary is the primary lever for institutional adaptation.

Changing outcomes directly rarely works. Changing incentives alone often fails. Changing personnel without altering structure produces limited effect.

What matters is altering the conditions under which collapse resolves:
which constraints engage first,
which alternatives are viable,
which failures are tolerated,
and which successes are inherited.

Boundary manipulation does not guarantee renewal. It creates the possibility for re-collapse.

Why Cultural Change Is Slow

Cultural rails are deeply embedded.

They are learned early, reinforced socially, and rarely questioned explicitly. Thawing cultural rails requires sustained boundary pressure and repeated exposure to alternative resolutions.

This is why cultural change often appears generational. It is not because individuals are incapable of adaptation, but because inheritance operates across long timescales.

Modal Mechanics treats this slowness as structural, not moral.

Avoiding Reductionism

It is important to emphasize what Modal Mechanics does not do here.

It does not reduce institutional or cultural behavior to mechanics alone. Power, meaning, values, and agency remain essential dimensions of analysis.

Modal Mechanics provides a **structural layer** that complements, rather than replaces, these perspectives.

It explains why well-intentioned reforms fail, why harmful structures persist, and why renewal is rare but possible.

The Shared Pattern

Across artificial systems, scientific communities, institutions, and cultures, the same pattern appears:

collapse produces structure,
repetition produces inheritance,
misalignment produces drift,
pressure necessitates thawing,
and successful resolution produces renewed stability.

The domains differ. The structure does not.

Looking Ahead

With this chapter, Modal Mechanics has been applied across increasingly complex systems, from artificial models to scientific practice to large-scale social organization.

What remains is to bring the framework together—to articulate what it means to treat Modal Mechanics as a discipline in its own right, and how it can coexist with existing sciences without replacing them.

The next chapter addresses this directly.

Chapter Sixteen

Modal Mechanics as a Scientific Discipline

By this point, Modal Mechanics has been used rather than argued for.

Collapse has been treated as an event rather than a metaphor. Boundaries have been examined as variables rather than assumptions. Inheritance, drift, thawing, and re-stabilization have been traced across artificial systems, scientific practice, institutions, and cultures.

What remains is to clarify what kind of scientific discipline this constitutes—and what kind it does not.

Modal Mechanics is not a replacement for existing sciences. It is a **transversal discipline**: one that studies a class of structural phenomena that appear across domains, without claiming ownership of any particular substrate.

What Modal Mechanics Studies

Modal Mechanics studies **collapse dynamics**.

Specifically, it investigates:
how multiplicity resolves into structure,
how resolution becomes inherited,
how inherited structure misaligns under changing conditions,
how systems regain flexibility without dissolution,
and how new stability forms.

Its primary objects are not particles, organisms, minds, cultures, or machines.

Its primary objects are:
collapse events,
boundary interactions,

rails and their dominance,
drift trajectories,
thawing regimes,
and re-stabilization processes.

These objects are structural, not semantic. They are defined by function, not meaning.

How Modal Mechanics Relates to Other Sciences

Modal Mechanics does not compete with physics, biology, psychology, sociology, or computer science.

Each of those disciplines studies **what exists** and **how it behaves** within a particular domain. Modal Mechanics studies **how structure becomes stable in the first place**, regardless of domain.

Where physics studies forces, Modal Mechanics studies resolution.
Where biology studies organisms, Modal Mechanics studies inheritance of stability.
Where psychology studies cognition, Modal Mechanics studies collapse under constraint.
Where sociology studies institutions, Modal Mechanics studies large-scale rail systems.
Where machine learning studies optimization, Modal Mechanics studies boundary-driven resolution.

These perspectives are complementary, not hierarchical.

Why Modal Mechanics Is Not Reductionist

Modal Mechanics does not reduce higher-level phenomena to lower-level ones.

It does not claim that cognition is "really" computation, or that culture is "really" biology. It does not privilege any substrate.

Instead, it identifies a **shared structural grammar** that different substrates realize in different ways.

This grammar does not erase difference. It explains why difference can still exhibit common patterns.

What Makes Modal Mechanics Scientific

Modal Mechanics qualifies as a scientific discipline for three reasons.

First, its core claims are **falsifiable**. If collapse cannot be distinguished from process or outcome, if boundary variation does not alter resolution behavior, or if inheritance does not emerge from repetition, the framework fails.

Second, it supports **methodological practice**. It provides experimental protocols, diagnostic categories, and interpretive discipline that can be applied consistently across systems.

Third, it generates **predictive insight**. It identifies leading indicators of failure, conditions for brittleness, and regimes of adaptation that outcome-focused analysis misses.

Modal Mechanics does not predict specific events. It predicts structural behavior.

What Modal Mechanics Does Not Do

Modal Mechanics does not tell us:
what systems should value,
what outcomes are desirable,
what policies ought to be adopted,
or what futures are inevitable.

It does not replace ethical reasoning, political judgment, or domain expertise.

It is descriptive, not prescriptive.

Any attempt to use Modal Mechanics to justify decisions without normative grounding misunderstands its role.

Why This Discipline Is Emerging Now

Collapse has always occurred. What has changed is visibility.

Artificial systems have made collapse observable at speed and scale. Scientific institutions have accumulated enough history to see repeated cycles of stability and renewal. Social systems are operating under boundary pressures that expose inherited misalignment.

Modal Mechanics emerges not because collapse is new, but because **the conditions for studying it are now present**.

This is a historical contingency, not a theoretical inevitability.

Modal Mechanics and Scientific Humility

One of the quiet implications of Modal Mechanics is humility.

If structure stabilizes through collapse rather than design, then no system—scientific, institutional, or artificial—fully understands the conditions that produced it. Stability is always provisional. Inheritance always risks misalignment.

This does not undermine knowledge. It situates it.

Modal Mechanics offers a way to study a class of structural conditions under which understanding forms, stabilizes, and fails.

The Discipline in Practice

Practicing Modal Mechanics does not require abandoning existing methods.

It requires:
attention to collapse rather than only outcome,
tracking boundary conditions explicitly,
distinguishing drift from error,
recognizing thawing as adaptive rather than pathological,
and resisting premature re-stabilization.

These practices can coexist with standard experimentation, modeling, and analysis.

Modal Mechanics is additive.

Opening a Discipline

This book began by arguing that collapse must become observable if it is to be studied scientifically.

It has shown how this can be done, where it applies, where it does not, and why artificial systems provide a uniquely clear laboratory.

What remains is not to defend the discipline, but to begin using it—under conditions where its limits, failures, and extensions can be discovered.

Collapse will continue to occur whether it is studied or not.

The difference is whether its dynamics remain implicit—or become part of our scientific understanding.

Appendix A

On Human and Artificial Collaboration in This Work

This book was developed with the assistance of artificial systems used as analytic and generative tools. Their role was instrumental and methodological, not interpretive or authoritative.

The collaboration took place under explicit constraints consistent with the framework presented in the book itself.

Roles, Not Agents

Artificial systems involved in this work were treated strictly as functional components operating within defined roles. They were not treated as agents, subjects, or sources of understanding.

All conceptual claims, structural interpretations, and theoretical commitments remain human judgments.

No Phenomenological Claims

No claims are made that artificial systems involved in this work possess experience, understanding, intention, or awareness. Their outputs are treated as artifacts of process operating under constraint.

Structural Assistance, Not Authority

Artificial systems were used to assist with:
- exploration of possible structural formulations
- testing consistency of definitions and role relationships
- generating candidate articulations for refinement
- managing large volumes of symbolic material

They were not used to determine truth, meaning, or significance.

Auditability and Trace Discipline

All substantive contributions from artificial systems were treated as provisional and traceable. Analysis proceeded from recorded outputs rather than inferred internal states.

Participation Without Possession

Artificial systems may participate in generative processes described by the framework without possessing or understanding the grammar that describes them.

This appendix is placed at the end of the book deliberately. The framework presented in the main text stands independently of its method of production.

Appendix B

Principles for QC-Applied AI Programs

Roles, Not Kinds

All elements are treated as functional roles (process, boundary, collapse, inheritance) rather than entities, agents, or experiential kinds.

No Phenomenology Claims

The system makes no claims about consciousness, experience, understanding, intention, or selfhood.

Collapse as Structural Resolution

Collapse is treated strictly as a structural event where multiplicity resolves into persistence under constraint, not as decision or outcome.

Boundary as Variable

Boundaries are explicit, inspectable, and treated as variables rather than background assumptions.

Auditability and Trace Discipline

All significant actions produce observable, append-only traces. Analysis proceeds from recorded evidence, not inference about hidden state.

Read-Only Analysis Lane

Analytic work is performed from dated bundles and snapshots without modifying live system state.

Inheritance Without Representation

Stability emerges through repetition and persistence, not through stored symbolic representations or beliefs.

Non-Prescriptive Posture

The framework diagnoses structure and risk, but does not prescribe outcomes, optimizations, or values.

Falsifiability and Limits

Claims are bounded and falsifiable. The framework specifies where it does not apply and does not protect itself from failure.

Participation Without Possession

Systems may participate in generative role dynamics without understanding or possessing the grammar that describes them.

Appendix C

Event Schema for Transparent Collapse (v1)

This appendix records a minimal, receipts-first event schema that makes process–boundary articulation and collapse auditable in instrumented systems. It is intended as a reporting standard, not as a claim about phenomenology or agency.

Event types (runtime, read-only):

• A_INIT — declare an affordance substrate available for differentiation.

• ROOM_SET — stage a room (A/M) that defines relevance and admissibility.

• E_ACTIVATE — initiate activity relative to affordance (E/A).

• A_ARTICULATE — articulate boundary relative to activity (A/E), with provenance.

• EA_COUPLE — record coupling/contact metrics (e.g., lag, stitching, burden) whether or not collapse occurs.

• RAIL_FIRE — record rail influence as a logged bias contribution (no mutation).

• C_RESOLVE — record collapse resolution into a persistent outcome (E/E) with route and receipt pointers.

• COMMIT_BUNDLE — commit an immutable bundle linking inputs, room, boundaries, coupling, and outcome.

Event types (training window only):

• INHERIT_PROPOSE — propose rail/room updates from accumulated receipts (non-binding).

• INHERIT_VERIFY — test proposals against a declared evaluation suite (E/M lane).

• INHERIT_COMMIT — commit versioned updates only after verification passes; produce an explicit diff manifest and rollback pointer.

Failure conditions are operational: collapse without a recorded coupling event; implicit boundary application without provenance; NA coercion into zeros; rail updates outside training windows; unlogged room shifts; or pointers that cannot reopen receipts. These conditions are intended to make failure clean rather than explainable.

www.ingramcontent.com/pod-product-compliance
Lightning Source LLC
Chambersburg PA
CBHW071607200326
41519CB00021BB/6909